MY LIFE AS A BRIDGE

I Am the Hawthorne Bridge in Portland Oregon

As Translated by Mark Parsons

MY LIFE AS A BRIDGE

I'm not the oldest bridge in Portland, nor am I the most beautiful. Those achievements would fall on other bridges in Portland as there are many. My nearest neighbor down-stream, the Morrison Street Bridge, could claim the oldest location for a bridge in Portland. The original Morrison Bridges before me were two. One was a wooden truss swing-span bridge completed on April 12, 1887. As the first Willamette River bridge in Portland and the longest bridge west of the Mississippi River, it can stand older in age than my location, even though it no longer exists.

I've lived my life as a bridge for more than one hundred years and there are no plans, that I know of, to replace me. I've served my community for all these years, first in their horse-drawn carts, chariots, hand carts and afoot, into the automobile age with their ever-increasing loads and abuse. Like all bridges before me, I know my purpose: to let pedestrians and traffic travel from one side of the Willamette River to the other and I've served that purpose well.

Being as old as I am, I thought it time to let you see the history that has happened around me these many years. Since photography was invented and developed shortly before I was built, I've decided to show those photographs that were taken from my superstructure or approaches. These photographs show what I've seen through the many years. In my one hundred plus years my towers and lift have been used to document the history of this part of Portland through these photographs. There will be many more photographs in my future years, I assume, if the city finds it feasible to leave me in place.

I love historic pictures, especially of the place I call home. Portland is a very young city with a very recent and very present history. I remember watching the log rafts float on the Willamette beneath me when I was newly built, and though I realize I am comfortably middle aged now, it feels not so long ago to me in years but so distant in memory.

My story begins not with me but with those first two bridges that I replaced, known as The Madison Street Bridges, or Madison Bridges. They refer to the two different bridges that spanned the Willamette River in Portland, Oregon from this very same location, from 1891 to 1900 and from 1900 to 1909. Those bridges connected Madison Street, on the river's west bank, and Hawthorne Avenue, on the east bank, on approximately the same alignment as where I sit today. The original and later bridges are sometimes referred to as Madison Street Bridge No. 1 and Madison Street Bridge No. 2, respectively. The second bridge, built in 1900, has alternatively been referred to as the "rebuilt" Madison Street Bridge (of 1891), rather than as a new bridge, because it was rebuilt on the same piers. Both were swing bridges, whereas I'm a vertical-lift-type. I'll explain that in a little while.

The second bridge in March 1908, when flooding upriver had caused a log jam to accumulate around it. The swing span is out of frame to the left in this view.

Construction of the first bridge, a wooden swing-span bridge, began in February 1890. It was built by the Pacific Bridge Company and owned by the Madison Street Bridge Company. It opened as a toll bridge on January 11, 1891. At that time, the bridge's east end was in the city of East Portland, Oregon. Subsequently, in July of the same year, East Portland merged with its larger neighbor, West Portland, becoming part of the City of Portland. Later in 1891, the Oregon state legislature organized eight Portland residents into a committee that purchased the bridge on November 18, 1891, for $145,000 (equivalent to $3,825,315 in 2015) and eliminated the tolls. The following year, the committee won approval from the United States Secretary of War for a contract to build the Burnside

Bridge, just down-river from my location, the second crossing of the Willamette River.

That bridge's two-lane roadway was 22 feet wide, and there were 6.5-foot sidewalks on both sides, while the structure's overall width was 40 feet.

Sorry to say, a disaster occurred on November 1, 1893 with this first bridge, when a westbound streetcar drove off the open draw of the bridge, and seven people died. This event remains the worst streetcar accident to occur in Portland, as well as the worst bridge disaster in the city's history. Nothing can make a bridge sadder than when someone dies in their attempt to just cross over to the other side.

This photograph, taken from this bridge shows the great Portland flood of 1894

In July 1899, that aging bridge was declared unsafe and in urgent need of rebuilding. Work to replace the structure, on the same piers, began in December 1899, with dismantling of the trusses.

In 1900 the first bridge was completely replaced by another wooden swing-span bridge, referred to as Madison Street Bridge No. 2. It has also been described as a rebuilding of the original bridge, because the work consisted of new truss spans mounted on the same piers. The rebuilt bridge opened to traffic in April 1900 but retaining the original swing-span section from 1891. Because of the estimated high cost, replacement of the swing span had been postponed, with predictions that it might hold on for another year or two. However, in July 1900 it was declared unsafe by the Multnomah County Board of Commissioners and plans to replace it with a new swing-type draw span were accelerated. The new swing span was constructed in fall 1900, and it was made stronger than its predecessor by an additional tower built over the middle pier, connected to the outer ends of the span by iron and steel rods.

This dramatic photo taken from bridge #1 in 1898 shows a fire in progress that destroyed at least a block of buildings along the Portland waterfront. We are looking west towards downtown from my north walkway. The taller building on the right is the U.S. Mills building owned by Albers and Schneider.

We bridges always worry about fire. Our steel structures can weaken and twist in a hot fire and a wooden bridge, of course, can burn completely. Two fires near my structure happened in 1898 (*Prior Photo*) and in 1902.

On June 21, 1902, a fire that destroyed six blocks of east-side waterfront property also heavily damaged the second bridge's eastern approach viaduct, closing the bridge to all traffic for several weeks. It reopened to pedestrians on July 18, to horse-drawn vehicles on August 5, and subsequently to streetcars.

By 1907, planning was under way for that bridge to be replaced by me, a new structure that would be positioned at a higher elevation over the water and be constructed of steel instead of wood. In June 1907, voters approved a measure to issue $450,000 in municipal bonds to fund construction of yours truly, me.

On January 20, 1909, the old number 2 bridge was closed indefinitely to all traffic, after high river levels had caused debris to accumulate around its piers, placing strain on the structure. The indefinite suspension became a permanent one. Plans for a new, stronger bridge, me, eventually to be named the Hawthorne Bridge, were firm by this time. The Portland Railway, Light and Power Company, whose streetcars had used the bridge, and many residents of the east side, argued that the old bridge should be reopened while the new one was being built, but it remained closed. Construction contracts for me were signed in June 1909. Dismantling of the old Madison Street Bridge's structure and demolition of its piers took place in August and September 1909, followed by my construction on the same alignment. I opened to pedestrian traffic first and wagon traffic later in the day on December 19, 1910.

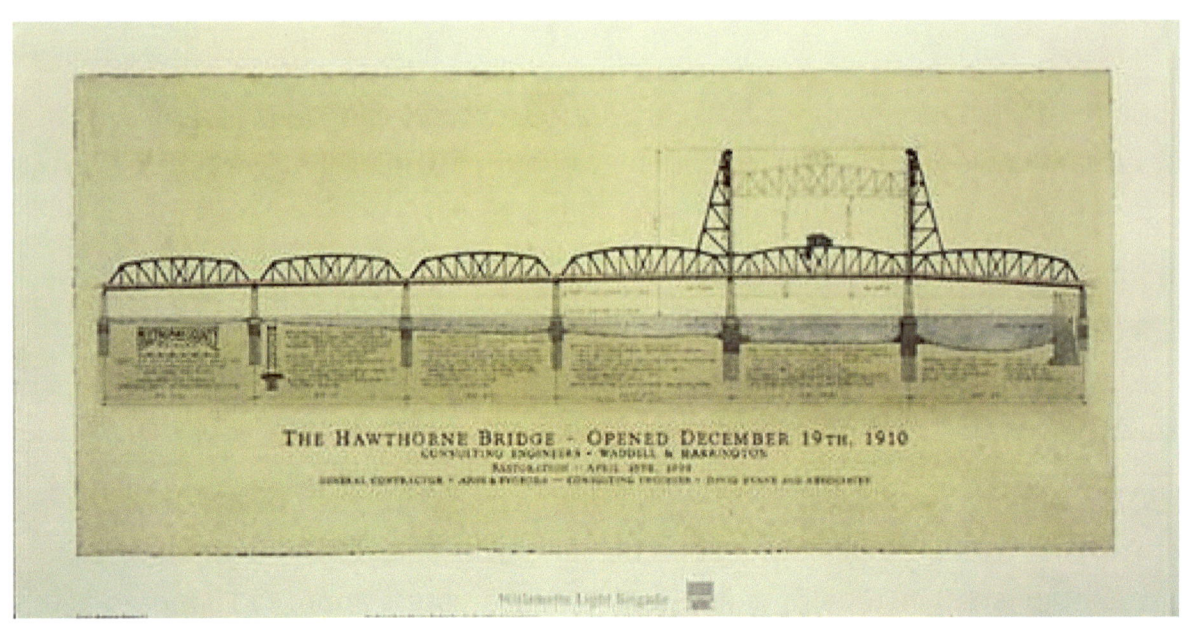

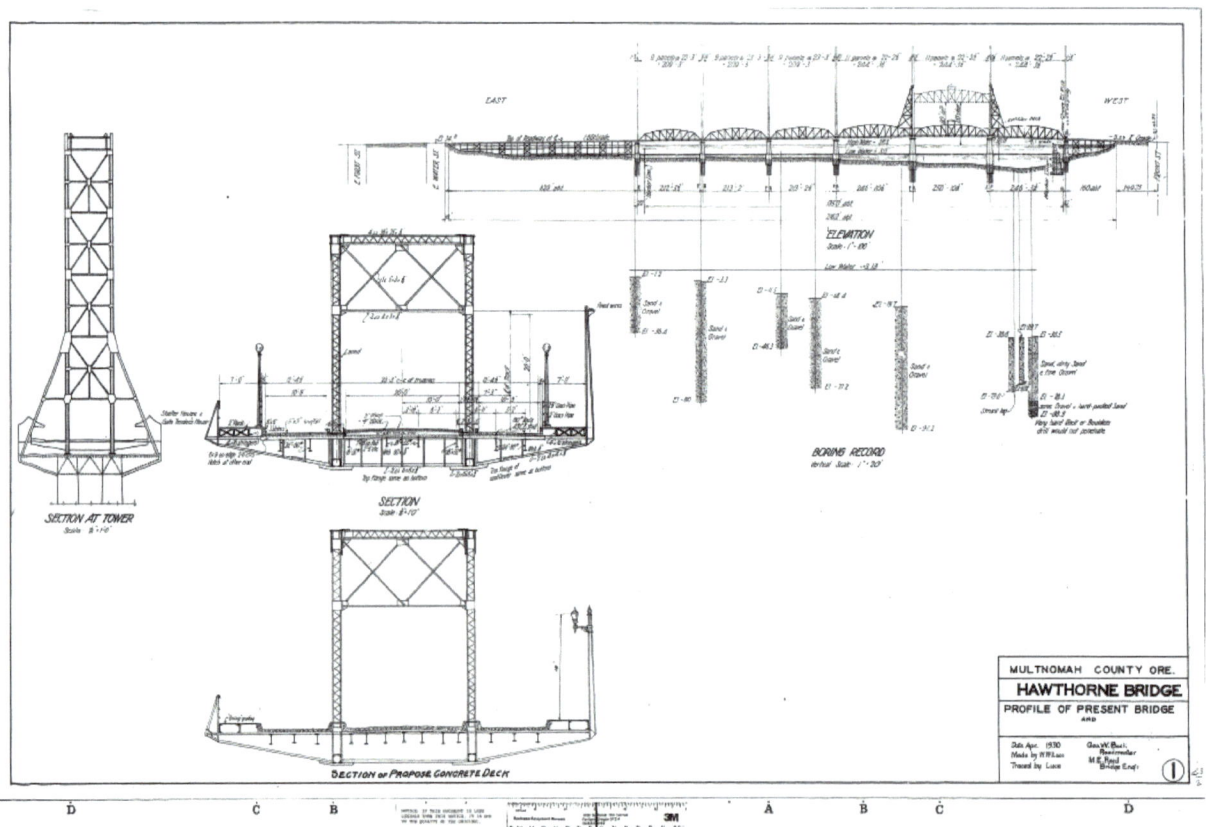

I was named for Hawthorne Boulevard, which I connected to on the east side, which was named after Dr. J.C. Hawthorne, the cofounder of Oregon's first mental hospital and early proponent for the first Morrison Bridge. My west end, of course, ended on Madison street.

I'm what's called a Parker Truss bridge with a vertical lift span that raises for ship traffic below. I have six trusses, one on my west end, the one which lifts and four on the east side of the lift. These are the iron workers who assembled my pieces:

Ironworkers assemble my structural framework in accordance with engineered drawings and they also install the metal support pieces for new buildings, as well as me. They also repair and renovate old structures using reinforced concrete and steel. Ironworkers may work on factories, steel mills, and utility plants. The work can be complicated.

This photograph shows the detail of my lift mechanism.

But enough of that detail stuff.

The streetcar tracks across my bridge were originally in the outer lanes but were relocated to the center lanes in 1931. The deck was changed from wood to steel grating in 1945.

My structure has been many colors over the years.

This 1993 photo also shows my original, narrower sidewalks.

In 1985 my lift span sheaves, the grooved wheels that guide my counterweight cables, were replaced. I went through a $21 million renovation from 1998–99, which included replacing the steel grated deck and repainting. My original lead-based paint was completely removed and replaced with 3 layers of new paint that is estimated to last 30 years. During this upgrade the sidewalks were widened to 10 feet, making it a thoroughfare for bicycle commuters. Due to the replacement of my steel deck during this project, my channels which used to carry the rails for streetcars and interurban trains were also removed. I was closed for one year to permit the renovation to be carried out. I needed the rest.

My original color was black, which lasted until 1964, when I was repainted yellow ochre. During the 1998–99 renovation, my color was changed to green with red trim, which I much prefer.

In 2001, my sidewalks were connected to the East bank Esplanade. In 2005, the city wanted an estimated cost to replace me. It totaled $189.3 million and I'm still here. As of 2001, my average daily traffic was 30,500 vehicles and counting.

I was added to the National Register of Historic Places in November 2012.

This 1993 photo also shows my original, narrower sidewalks.

On the next page will be found a bird's eye view drawing of the Portland's bridges in 1890.

Let's look at some of the photographs that have been taken from my structure:

Sometimes the photographer would even colorize the same photographs.

Or ad objects, like these sternwheelers.

Photographs taken from my northern walkway

Stern wheeler backing towards me

This photo, taken from my predecessor's traffic deck in 1901

From my tower in 1921

Showing repair or construction of waterfront pilings

Another from my deck shows wharves closer to my west end about 1928. That's one of the Morrison Bridges in the background

Although not from my deck, I thought you might like to see this photograph from just before the seawall was constructed in 1928-9

And another

Navy ships before 1929

USS Oregon (BB-3) the third and final member of the Indiana class of pre-dreadnought battleships built for the United States Navy in the 1890s.

The ship was decommissioned in 1919 and efforts by naval enthusiasts in the early 1920s led the Navy to loan Oregon to her namesake state for use as a museum ship. Shown in 1939 from my southern deck during her time as a museum on the southern end of the seawall.

After the start of World War II, the Navy decided in late 1942 to scrap the ship for the war effort, but after work began the Navy requested the ship's return for use as an ammunition hulk for the upcoming invasion of Guam in 1944.

The Portland Public Market built in 1933.

Three stories tall with eleven-story towers, three blocks long, and with features including a gas station, rooftop parking, and a 500-seat auditorium, it was primarily a novelty, and struggled to retain tenants until finally closing in 1942. The architect was William G. Holford.

The building was leased to the U.S. Navy in 1943, then sold to The Oregon Journal in July 1946, for use as the newspaper's operations plant starting in 1948. After publishing from there for 13 years, the paper moved out in 1961, and the building stood unused until it was bought in 1968 by the City of Portland, which demolished it the next year to make way for an expansion of Harbor Drive, which itself was largely replaced in 1974 by Tom McCall Waterfront Park.

But over the years it was photographed many times, usually for whatever was happening between myself and it.

I love water and ships, so I'm also going to include another photograph not taken from by deck. This one was taken from my sister-bridge Morrison

This is another photo, but taken from my east approach.

The waterfront roadway was eventually removed to make way for Tom McCall Waterfront Park.

Flood of 1948

*My East end ramparts, or approaches and exits were built after 1957.
They were built over the roads and railroad tracks, which were a hinderance to traffic into and out of downtown Portland and over southeast Madison and southeast Hawthorne Boulevard*

East side approaches still under construction.

Site of the new Multnomah County Building on my west end approach, taken from a Livestream video atop my southwest lift tower.

Above, A later view

Since 1974 the Tom McCall Waterfront Park was built along and under my structure, down to where the battleship Oregon once was moored as a floating museum. I've seen the Oregon Museum of Science and Industry being built where the old Station L Power Plant was donated by Portland General Electric company. In my time I've seen transportation developed from wagon to model T's to the sleek new automobiles of today.

The Marquam Bridge is a marvel of improvement over my construction. It's a double-deck, steel-truss cantilever bridge that carries Interstate 5 traffic across the Willamette River south of my position on the Willamette. It is the busiest bridge in Oregon, carrying 140,500 vehicles a day as of 2016. I'm impressed.

This is Fire Station 21 on the east bank of the Willamette River from the east end of my walkway. I remember when this was docks and ship ways or repairs, warehouses and railroad tracks. It is now part of the Eastside walking trail now.

Above: Before and After.

There have also been some very bad photographs that were doctored to look like they were taken from my structure. Like this one:

That's the end of our little tour of historical photographs taken from my structure over the years. If you like it, let me know and my relatives up and down the river would also like to show you their photos.

www.ingramcontent.com/pod-product-compliance
Lightning Source LLC
Chambersburg PA
CBHW040440220526
45473CB00004B/1485